MATH FUN

Picture Window Books
Minneapolis, Minnesota

If you were... a plus sign, a minus sign, a times sign, a divided-by sign, an even number, an odd number, a fraction, a set

Editor: Christianne Jones and Jill Kalz
Designer: Lori Bye
Page Production: Melissa Kes
Art Director: Nathan Gassman
Editorial Director: Nick Healy
Creative Director: Joe Ewest
The illustrations in this book were created with acrylics.

Picture Window Books
151 Good Counsel Drive
P.O. Box 669
Mankato, MN 56002-0669
877-845-8392
www.picturewindowbooks.com

Copyright © 2010 by Picture Window Books
All rights reserved. No part of this book may be reproduced
without written permission from the
publisher. The publisher takes no responsibility
for the use of any of the materials or methods
described in this book, nor for the products thereof.

Printed in China.

Library of Congress Cataloging-in-Publication Data
Aboff, Marcie.
Math fun / written by Marcie Aboff
and Trisha Speed Shaskan; illustrated
by Francesca Carabelli and Sarah Dillard.
p.cm. — (Math fun)
Includes index.
ISBN 978-1-4048-5611-0 (paperback bindup)
1. Mathematics—Juvenile literature.
2. Arithmetic—Juvenile literature.
3. Mathematical notation—Juvenile literature.
4. Geometry—Juvenile literature.
I.Shaskan, Trisha Speed, 1973- II.
Carabelli, Francesca, ill. III. Dillard, Sarah, ill.
IV. Title.
QA40.5.A26 2010
510—dc22   2009007391

Special thanks to our advisers for their expertise:
Rosemary G. Palmer, Ph.D., Department of Literacy
College of Education, Boise State

Terry Flaherty, Ph.D., Professor of English
Minnesota State University, Mankato

Table of Contents

If you were ...
a plus sign.....................6
a minus sign...................28
a times sign............50
a divided-by sign.............72
an even number...................94
an odd number..................116
a fraction...........138
a set.....................160

Summary......................182
Glossary..................184
To Learn More.................186
Index.........................188

plus sign (+) a symbol used to show addition

If you were
a plus sign . . .

Ida and Ike have a cub called Ina.
They make a family of three.

Ida, Ike, and Ina love pets. They have two seals.
They make a family of five.

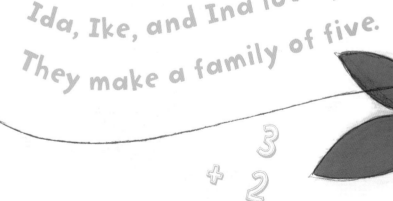

If you were a plus sign, you would be a symbol used to show addition. You would be part of an addition problem.

Selma juggles five red balls plus two green balls. That's a total of seven balls.

$$\begin{array}{r}5\\+2\\\hline 7\end{array}$$

If you were a plus sign, you would be used like the word "and."

Doodle gives Pluck three daisies and six roses.
Three and six equals nine.

3
+ 6
―――
9

If you were a plus sign, you would help make a sum. The sum is the total of two or more numbers.

5 + 5 = 10

Five bulldogs met five tigers to play basketball at the park. They made a sum of ten players.

Six bulldogs plus six tigers showed up to cheer them on. They made a sum of twelve cheerleaders.

$$\begin{array}{r}6\\+6\\\hline 12\end{array}$$

If you were a plus sign, you could work left to right or top to bottom.

4 + 3 = 7

Spotty cooks four blueberry pancakes for herself and three plain pancakes for Dotty.

Spotty stacks the three plain pancakes and the four blueberry pancakes on a plate.

$$\begin{array}{r}3\\+4\\\hline 7\end{array}$$

If you were a plus sign, you could add in any order and still get the same sum.

A frightened frog leaps across four lily pads and then bounces onto two more lily pads. He jumps on six lily pads altogether.

4+2=6

If you were a plus sign, you could add any amount of numbers. You could add two, three, four, or more. You would get the same sum no matter how the numbers were arranged.

Five elephants plus four hippos plus three rhinos equals twelve show animals.

5+4+3=12

Five elephants plus three rhinos plus four hippos equals twelve show animals.

If you were a plus sign, you could add small numbers or big numbers.

Gert wears eight red bracelets on her right arm and eight pink bracelets on her left arm. Gert is wearing sixteen bracelets.

8+8=16

Gert has thirteen bows on her tail and fourteen bows in her mane. She's wearing twenty-seven bows.

13+14=27

Gert has 100 polka dots on her skirt, and Stretch has 100 triangles on his vest and tie. Together, they have two hundred shapes on their clothes.

100+100=200

If you were a plus sign, you could help solve story problems.

Cherry juggles five red balls. She adds three green balls and two orange balls. How many balls does Cherry juggle in all?

5+3+2=10

Cherry juggles ten balls in all.

You would always help add things together ...

... if you were a plus sign.

If you were a minus sign, you would be a symbol used to show subtraction. You would be part of a subtraction problem.

Starry-eyed Stella's daisy has nine petals.
Stella plucks eight petals from it.
Stella's daisy has one petal left.

If you were a minus sign, you would be used in place of the words "take away."

Mama Munk finds twelve acorns.
She takes away five to hide.

Twelve take away five is seven.
Seven acorns are left.

12
− 5
―――
7

Papa Munk finds seven acorns.
He takes away five to hide.
Seven take away five is two.
Two acorns are left.

$7 - 5 = 2$

If you were a minus sign, you would help show the difference in a subtraction problem.

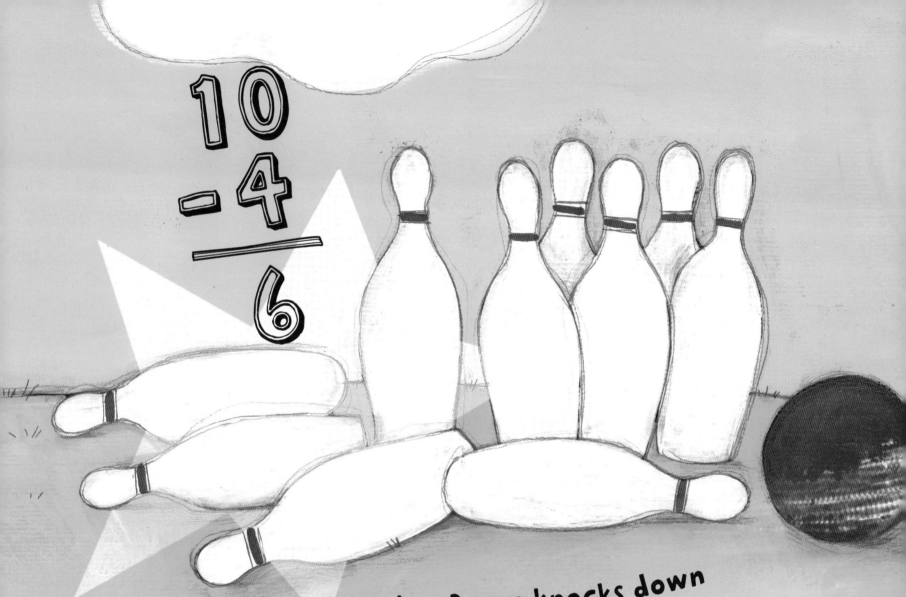

There are ten bowling pins. Bruno knocks down four of them. Six bowling pins are left standing. The difference between ten and four is six.

If you were a minus sign, you could work left to right or top to bottom.

Patty fries seven burgers. She puts five on a plate. Two are left in the pan.

7 - 5 = 2

If you were a minus sign, you could help solve a story problem.

Sandy blew eleven bubbles.
Gill popped four of them.
How many bubbles were left?
Eleven minus four equals seven.
Seven bubbles were left.

If you were a minus sign, you could subtract small numbers.

There are nine glasses of fruit punch on the table. Three thirsty tigers take three glasses away. There are six left.

9−3=6

Jump into SUBTRACTION

Using sidewalk chalk, draw a large number line. A number line looks like a ruler. Then draw marks on it for each number. Write the numbers in order from one to ten. Leave enough space between each number to jump from one to the next. Now, it's time to subtract.

1. Start on the number ten. Jump three numbers down the line toward the number one. The number you landed on is the difference between ten and three. 10-3=?

2. Start on the number eight. Jump four numbers down the line toward number one. The number you landed on is the difference between eight and four. 8-4=?

3. Start on the number seven. Jump three numbers down the line toward number one. The number you landed on is the difference between seven and three. 7-3=?

Now make up at least five subtraction problems of your own. Solve them by jumping down the number line!

Answers: 1. (7) 2. (4) 3. (4)

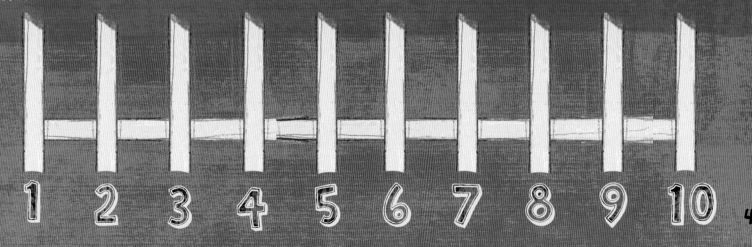

times sign (x) a symbol used to show multiplication

50

Five sneaky spiders have eight legs each. Five spiders times eight legs each equals forty legs in all.

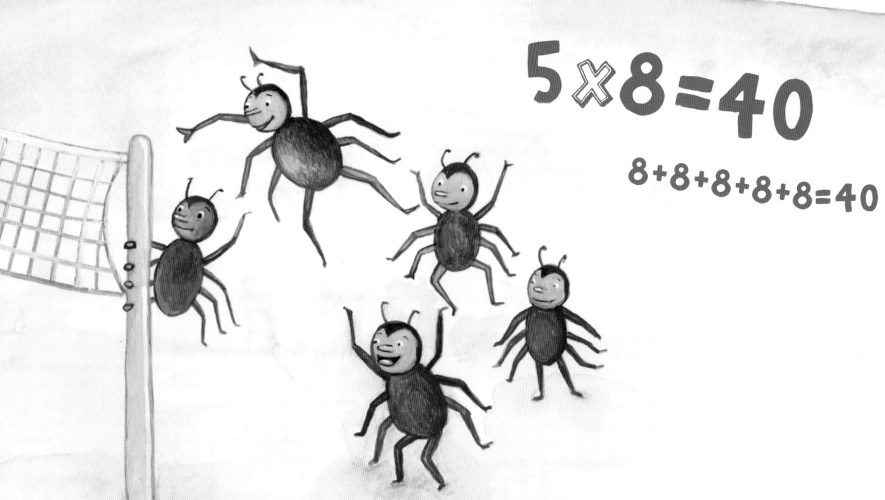

$5 \times 8 = 40$

$8+8+8+8+8=40$

If you were a times sign, you would be part of a multiplication problem. The problem's answer is called the product.

Ribbon snakes have three white stripes on their backs. Three ribbon snakes times three white stripes each equals nine white stripes total. **The product is nine.**

3 × 3 = 9

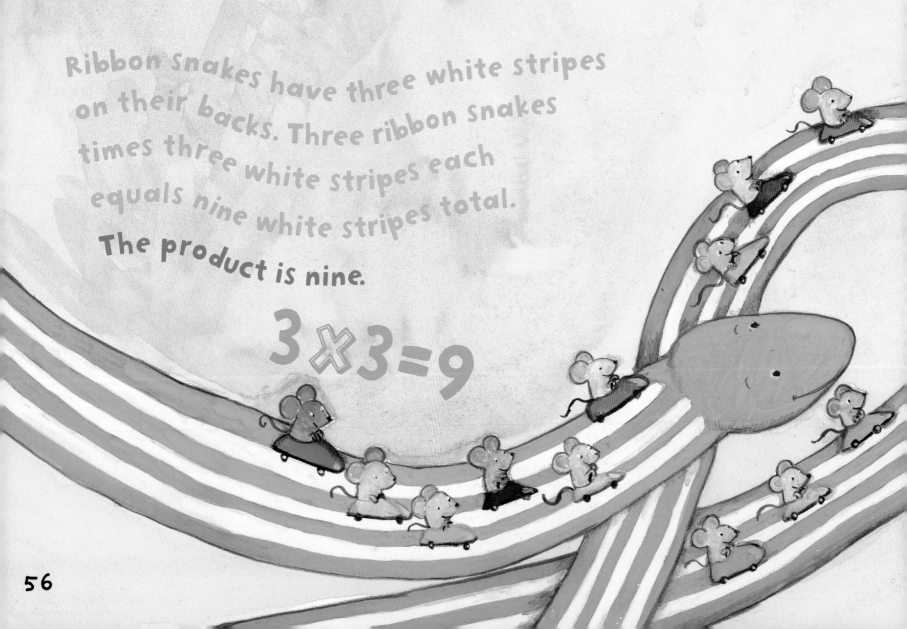

Two mice race on each white stripe. Two mice times nine stripes equals eighteen mice total. The product is eighteen.

2 ✕ 9 = 18

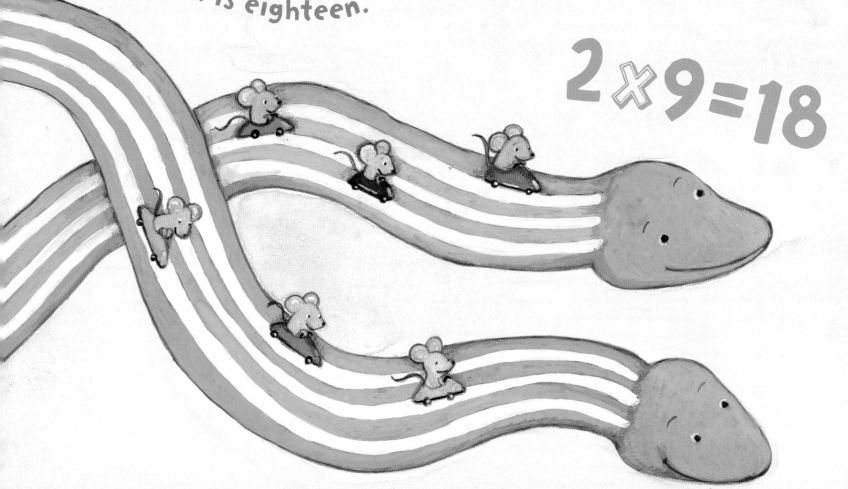

If you were a times sign, you could switch the number before you and the number after you. It wouldn't matter which number came first because the product would be the same.

2×3=6

Two tigers' tricycles have three wheels each. Two tricycles times three wheels each equals six wheels total.

Three bears' bicycles have two wheels each. Three bicycles times two wheels each equals six wheels total.

$3 \times 2 = 6$

If you were a times sign, you could work left to right or top to bottom.

Toucan has two toasters.
Each toaster holds four pieces of bread.
Toucan fills the two toasters.

Two toasters times four pieces of bread each equals eight pieces of bread in all.

2×4=8

Toucan takes the toast to the twins. Two plates times four pieces of toast on each equals eight pieces of toast in all.

If you were a times sign, you could be used like the words "groups of."

Fifi holds two groups of five balloons.
Two groups of five equals ten balloons.

$$\begin{array}{r} 2 \\ \times 5 \\ \hline 10 \end{array}$$

$2 \times 5 = 10$

Suddenly, the flippers slip off! Twenty-four lost flippers divided by eight empty arms on each octopus equals three sad octopuses.

$24 \div 8 = 3$

$24 - 8 - 8 - 8 = 0$

If you were a times sign, you could multiply single-digit or double-digit numbers.

Two seven-eyed aliens land on Saturn. Two aliens times seven eyes each equals fourteen eyes total.

2 ✕ 7 = 14

Ten ten-eyed aliens greet them. Ten aliens times ten eyes each equals 100 eyes total.

$$\begin{array}{r} 10 \\ \times 10 \\ \hline 100 \end{array}$$

You would always multiply ...

Multiplication Table

×	0	1	2	3	4	5	6	7	8	9	10
0	0	0	0	0	0	0	0	0	0	0	0
1	0	1	2	3	4	5	6	7	8	9	10
2	0	2	4	6	8	10	12	14	16	18	20
3	0	3	6	9	12	15	18	21	24	27	30
4	0	4	8	12	16	20	24	28	32	36	40
5	0	5	10	15	20	25	30	35	40	45	50
6	0	6	12	18	24	30	36	42	48	54	60
7	0	7	14	21	28	35	42	49	56	63	70
8	0	8	16	24	32	40	48	56	64	72	80
9	0	9	18	27	36	45	54	63	72	81	90
10	0	10	20	30	40	50	60	70	80	90	100

... if you were a times sign.

MULTIPLICATION FUN

What you need:
an empty egg carton
50 pennies
a pen or pencil
a piece of paper

What you do:
1. Fill six of the cups in the egg carton with two pennies each.
2. Count the number of pennies. The total is the product of 6x2.
3. Fill six of the cups in the carton with three pennies each. Count the number of pennies. The total is the product of 6x3.
4. Create your own multiplication problems. For example, fill some of the cups with four pennies each. Be sure to use the same number of pennies in each cup. If you fill one cup with two pennies, fill them all with two pennies. Multiplication is just repeated addition!
5. List the multiplication problems and the products you come up with on a piece of paper. Use the multiplication table on page 70 to check your answers!

divided-by sign (÷) a symbol used to show division

If you were a divided-by sign ...

73

... you would divide one number by another number.

Four hungry hyenas have a pizza party. They divide a large pizza equally. Sixteen slices divided by four hyenas equals four slices each.

16 ÷ 4 = 4

If you were a divided-by sign, you would be a symbol used to show division. Division shows how many times one number goes into another number.

It's the first wrestling match of the season! Eight tangled legs divided by two wrestlers equals four legs each.

8 ÷ 2 = 4

Two goes into eight four times.

Twelve clapping paws divided by six cheerleaders equals two clapping paws each. Six goes into twelve two times.

12 ÷ 6 = 2

If you were a divided-by sign, you would be part of a division problem. Division is a quick way to subtract the same number over and over.

Three penguins celebrate their birthdays. They receive twenty-one pretty presents and eighteen colorful balloons.

Twenty-one divided by three equals seven presents each.
Eighteen divided by three equals six balloons each.

$21 \div 3 = 7$

21-3-3-3-3-3-3-3=0

$18 \div 3 = 6$

18-3-3-3-3-3-3=0

If you were a divided-by sign, you would stand between two numbers. The number being divided is called the dividend. The number that divides it is called the divisor. The quotient is the answer to the division problem.

If you were an even number, you would be divisible by 2. You would never have a remainder.

A group of 8 animals wanted to board the roller coaster. The ride had 4 cars.

The 2 mice took the first car.

The 2 birds took the second car.

The 2 bears took the third car.

The 2 cats took the fourth car.

The 8 animals were evenly divided into 4 cars.
Both 8 and 4 are even numbers.
No one got left out.

If you were an even number, your last digit would be 0, 2, 4, 6, or 8.

The leaping lizards played with 10 jacks.

The funny frogs played with 26 marbles.

The puzzled penguins played with a deck of 52 cards.
Go fish!

If you were an even number, you would be between two odd numbers. You would be every other counting number on a number line, starting with 2.

Mrs. Rabbit asked the class to match the pictures on the board to the even numbers on the number line.

If you were an even number, you could add another even number to yourself. You would always get another even number.

The dog chased 2 balls.
Then he chased 4 more balls.
The dizzy dog chased 6 balls.

2+4=6

The cat climbed 8 branches.
Then he climbed 10 more branches.
The crazy cat climbed 18 branches.

If you were an even number, you could be an instrument and make beautiful music.

Marvin tapped 2 drumsticks.

Mary plucked 6 strings on a guitar.

Milt played 88 keys on a piano.
Melanie sang 8 sassy songs.

The 4 monkeys rocked the house!

If you were an even number, you would stand on two feet, four feet, or eight feet. You wouldn't stand on one foot, because that's odd!

The 8-footed spider helped the 2-footed penguin and the 4-footed polar bear pack to go and visit their friend, the flamingo.

If you were an even number, you could go to a twins' picnic.

Two by 2, all of the animals came to the twins' picnic. Two pigs cooked 12 hot dogs, 10 hamburgers, and 22 ears of corn.

FUN WITH EVEN NUMBERS

Gather as many pennies as you can find and answer the questions below.

1. How many pennies did you collect in all? Is it an even number?

2. Count the pennies by twos. See how high you can count. Are there any pennies left over?

3. Put all of the pennies in a cup. Toss the pennies on the floor. How many pennies have heads showing? Is it an even number? Now do the same thing counting tails.

4. See if you can stack the pennies without them falling over. Count them as you stack them. How many pennies can you stack before they fall over? Was it an even number?

odd number—a number that is not divisible by two

If you were an odd number...

... you could be the starting lineup.

The 5 basketball players scrambled down the court.

The 9 baseball players ran onto the field.

The 11 football players huddled by the end zone.

If you were an odd number, you would not be divisible by 2. There would always be a remainder of one left over.

The tree had 3 bananas. Two monkeys each ate 1 banana. One banana was left over. Three is an odd number.

Two mice had 7 pieces of cheese. Each mouse ate 3 pieces of cheese. One piece of cheese was left over. Seven is an odd number.

If you were an odd number, your last digit would be 1, 3, 5, 7, or 9.

Patrick guessed there were 253 jellybeans in the jar.

Paula guessed there were 459 jellybeans in the jar.

Patricia guessed there were 787 jellybeans in the jar.

Patsy guessed the right answer. There were 645 jellybeans in the jar!

But if you were an odd number and added yourself to another odd number, you would get an even answer.

Maggie gathered 7 (odd) green apples and 3 (odd) red apples. She had 10 (even) apples in all.

If you were an odd number, you could be the seven days of the week.

Monday, Tuesday, and Wednesday were rainy. It rained 3 days.

Thursday, Friday, and Saturday were snowy. It snowed 3 days.

Sunday was sunny. It was the 1 perfect day to throw snowballs!

If you were an odd number, you could light up the sky.

Seven lazy lobsters worked on their tans under 1 shining sun.

FUN WITH ODD NUMBERS

Sometimes it's cool to be odd. Find out how odd you are by answering the following questions. See how many odd numbers you get. Then ask your friends how odd they are.

1. How old are you?

2. What date is your birthday?

3. What is the last number in your telephone number?

4. What is the last number in your home address?

5. How many people are in your family?

fraction—one or more equal parts of a whole

If you were a fraction, you could be divided into three equal parts. You would be thirds.

Jenny juggles three red balls. One ball falls. Jenny has dropped one-third of the balls.

The proud peacock's flag is one-half green and one-half purple.

1/2

1/2

If you were a fraction, you could be divided into four equal parts. You would be fourths.

One window has four equal parts, or panes. If one pane is broken, one-fourth of the window is broken.

If three panes are broken, three-fourths of the window is broken.

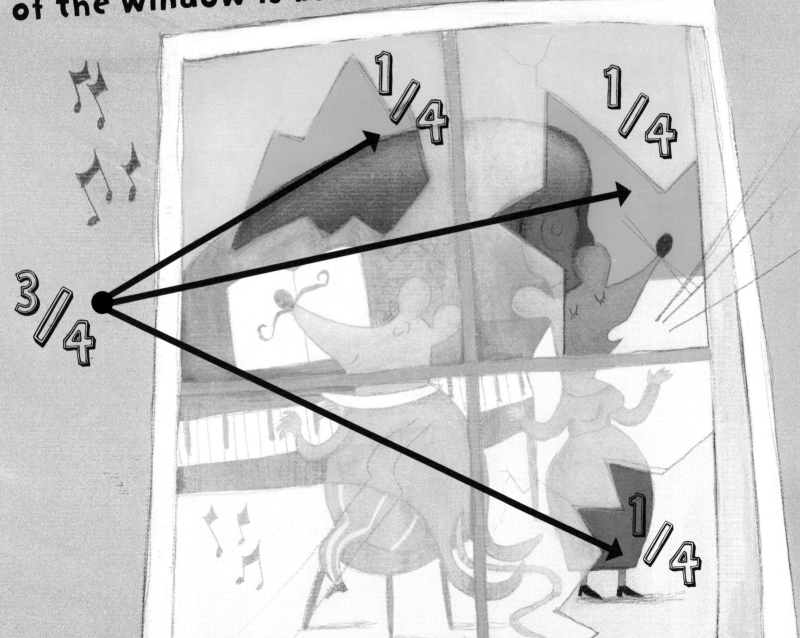

If you were a fraction, you could be divided into eight equal parts. You would be eighths.

Robin slices a raspberry pie into eight pieces. Each bird gets one piece. Each bird gets one-eighth of the pie.

If you were a fraction, you could be divided more than once. You would be one unit subdivided into equal parts.

Josie and Jackie split the yummy candy bar. They each have one-half.

Then Johnny and Joey show up. Josie and Jackie split each part again. Now the candy bar is in four pieces.

Josie, Jackie, Johnny, and Joey each have one-fourth of the candy bar.

If you were a fraction, you could be part of a set. A set is a group that has something in common.

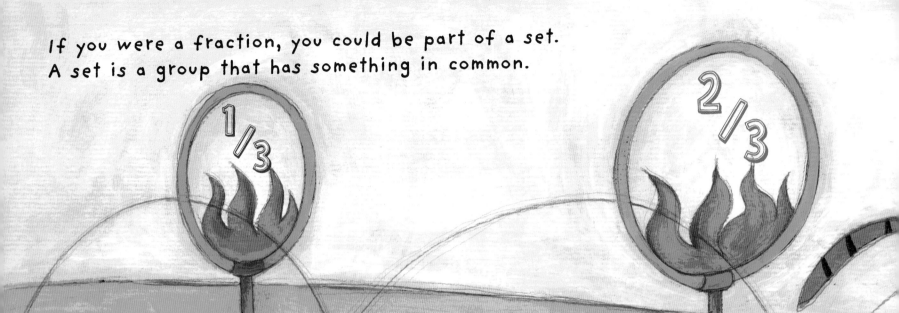

Louis the tiger trainer has a set of three flaming hoops. Fanny the tiger jumps through the first flaming hoop, or one-third of the set.

Fanny jumps through the second flaming hoop. Fanny has now jumped through two-thirds of the set.

Fanny jumps through the third flaming hoop. Now Fanny has jumped through three-thirds of the set, or one whole set.

If you were a fraction, you could be compared with other fractions.

Buck, Buttercup, and Betsy make six cookies.

Buttercup gets one of the six cookies. She has one-sixth of the cookies.

1/6

Buck gets two of the six cookies. He has two-sixths, or one-third, of the cookies.

2/6 (1/3)

Betsy gets three of the six cookies. She has three-sixths, or one-half, of the cookies.

3/6 (1/2)

One-sixth is the smallest of the three fractions. Buttercup has the fewest cookies. Three-sixths is the largest of the three fractions. Betsy has the most cookies.

You would always be part of the whole ...

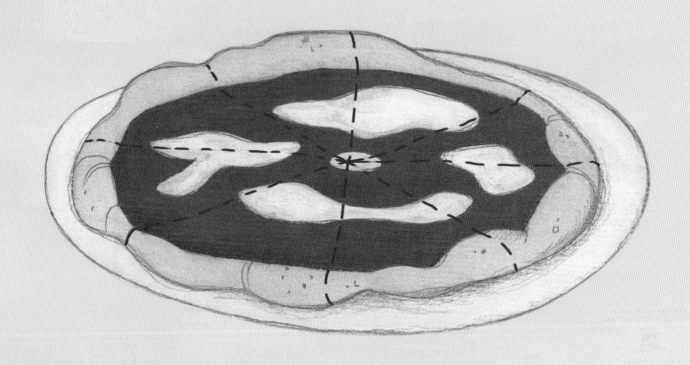

... if you were a fraction.

FRACTION FUN: Make your own pizza pie!

What you need:

- a circular object to trace, such as a coffee can, a Frisbee, or the plastic top from an oatmeal container
- a piece of paper
- scissors
- crayons or markers
- a pencil

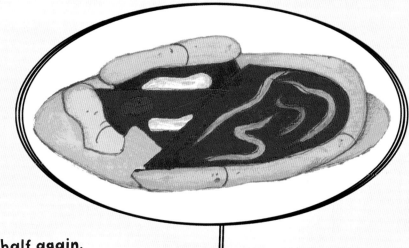

What you do:

1. Trace a circle onto a piece of paper.

2. Cut out the circle.

3. Fold the circle in half. Fold it in half again. Fold it in half again.
   (It will be the shape of an ice-cream cone.)

4. Unfold the paper. Use a pencil to trace the lines where the paper was folded. Your lines should create eight equal parts.

5. Use your crayons or markers to draw pizza toppings. You can draw whatever you want on each slice. Here are some ideas: pepperoni, mushrooms, pineapple, sausage, and green peppers. You may want to cover more than one slice with the same topping. You may want to put two toppings on one slice.

6. Now find some fractions in your pizza. If two slices have mushrooms and sausages on them, mushrooms and sausages cover 2/8 (or 1/4) of the pie. If three slices have green peppers on them, green peppers cover 3/8 of the pie. Make a list of the fractions you used.

If you were a set, you could be matched with objects in another set.

Mimi laid out a set of six colorful crayons.
Mo had a set of six pretty paints.
Mildred opened a set of six marvelous markers.

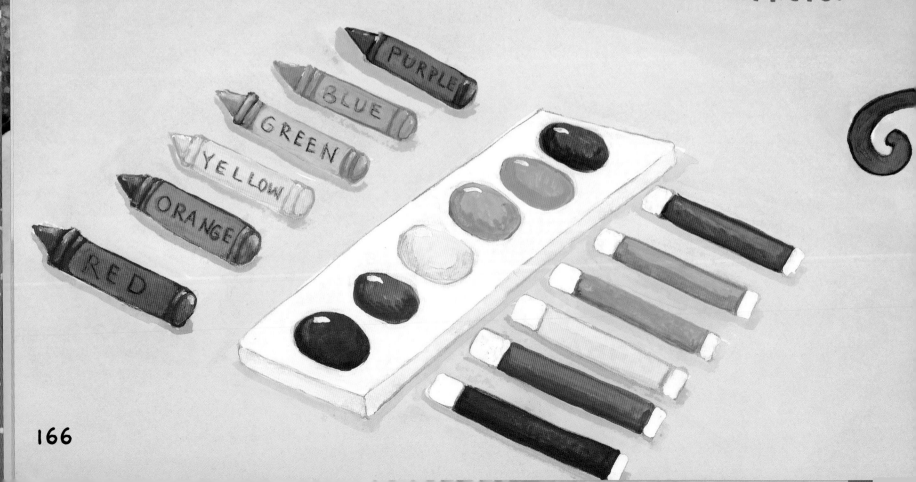

If you were a set, you could be arranged in different ways.

Lilly picked six peaches, two pineapples, and eight pears at the fruit stand.

She carried the fruits home, cleaned them, and created a fine fruit bowl with her set of fruits.

If you were a set, you could take away a subset and still be a set. The remaining set would also be a subset.

Sally wanted to pack five pairs of shoes. But she could fit only two pairs of shoes in her suitcase. The two pairs are a subset.

Now Sally has a subset of three pairs of shoes left in her closet. Her set of five pairs of shoes is separated into two subsets.

If you were a set, you could be a combination of other sets.

Betty built a set of two yellow towers.

Bobby came to play and built
a set of three blue towers.

Billy came to play and built
a set of four green towers.

If you combine the three sets together,
you get one set of nine towers.

If you were a set, you could be compared to other sets.

Johnny picked a set of six roses, a set of four daisies, and a set of four petunias.

The set of roses was a larger set than either the set of daisies or the set of petunias.

The sets of daisies and petunias together was a larger set than the set of roses.

The sets of daisies and petunias had the same size.

Johnny combined the sets and gave the beautiful bouquet to Josie. She was overjoyed!

If you were a set, you would always know what belongs in your group.

Proper Pat had a perfect place setting of polka-dot dishes. Oh, no! One plate fell!

Pat had to use a plate from another set. He was very upset that the plates were not all the same.

FUN WITH SETS

Sets are collections of things. Lots of people collect things such as toy cars, stuffed animals, stamps, and favorite books.

Make a collection of rocks. Start by collecting 10 rocks. Then sort them into subsets. Sort them by size, color, or shape.

Now make a collection of coins. Sort them by value (penny, nickel, dime, quarter). Then sort them by year.

What else can you collect?

SUMMARY

plus sign (+) a symbol used to show addition

minus sign (−) a symbol used to show subtraction

times sign (x) a symbol used to show multiplication

divided-by sign (÷) a symbol used to show division

even number—a number that is divisible by two

odd number—a number that is not divisible by two

fraction—one or more equal parts of a whole

set—a group that has something in common

GLOSSARY

add—to find the sum of two or more numbers

addition—the act of adding numbers together

arrange—to place in an order

difference—the number left after subtracting one number from another

digit—any of the numbers 1 through 9, and sometimes 0

divided—separated into parts or groups

divided-by sign—a symbol used to show division

dividend—the number being divided in a division problem

divisible—able to be separated into equal parts

division—how many times one number goes into another number, or a quick way to repeatedly subtract the same number

divisor—the number doing the dividing in a division problem

even number—a number that is divisible by two

fraction—one or more equal parts of a whole

minus sign—a symbol used to show subtraction

multiplication—a quick way to repeatedly add the same number

odd number—a number that is not divisible by two

plus sign—a symbol used to show addition

product—the number found from multiplying two or more numbers

quotient—the answer to a division problem
remainder—a leftover number
set—a group that has something in common
sort—to separate
subdivided—one thing divided more than once
subset—a set that is part of a larger set
subtract—to take away one part, or number, from another
subtraction—taking one number from another number
sum—the number you get when you add two or more numbers together
symbol—a sign that stands for something to be done
tally marks—marks that show the number of items; usually grouped by five
times sign—a symbol used to show multiplication

185

TO LEARN MORE

More Books to Read

Aber, Linda Williams. *Grandma's Button Box.* New York: Kane Press, 2002.

Bauer, David. *Adding Arctic Animals.* Bloomington, Minn.: Yellow Umbrella Books, 2004.

Chrismer, Melanie. *Multiply This!* New York: Children's Press, 2005.

Chrismer, Melanie. *Odd and Even Socks.* New York: Children's Press, 2005.

Cleary, Brian P. *The Action of Subtraction.* Minneapolis: Millbrook Press, 2006.

Cleary, Brian P. *The Mission of Addition.* Minneapolis: Millbrook Press, 2005.

Ekeland, Ivar. *The Cat in Numberland.* Chicago: Cricket Books, 2006.

Fisher, Doris, and Dani Sneed. *My Even Day.* Mount Pleasant, S.C.: Sylvan Dell Pub., 2007.

Fisher, Doris, and Dani Sneed. *One Odd Day.* Mount Pleasant, S.C.: Sylvan Dell Publishing, 2006.

Franco, Betsy. *Subtraction Fun.* Mankato, Minn.: Yellow Umbrella Books, 2002.

Freeman, Marcia S. *Multiply by Hand: The Nines Facts.* Vero Beach, Fla.: Rourke Pub., 2008.

Gisler, David. *Addition Annie.* New York: Children's Press, 2002.

Hall, Pamela. *The Odds Get Even!: The Day the Odd Numbers Went on Strike.* Los Angeles: Piggy Toes Press, 2003.

Jaffe, Elizabeth Dana. *Can You Eat a Fraction?* Mankato, Minn.: Yellow Umbrella Books, 2002.

Koomen, Michele. *Sets: Sorting into Groups.* Mankato, Minn.: Bridgestone Books, 2001.

Leedy, Loreen. *Subtraction Action.* New York: Holiday House, 2002.

Lewis, J. Patrick. *Arithme-Tickle: An Even Number of Odd Riddle-Rhymes.* San Diego: Harcourt, Inc., 2002.

Murphy, Stuart J. *Seaweed Soup.* New York: HarperCollins, 2001.

Nagda, Ann Whitehead. *Cheetah Math: Learning About Division from Baby Cheetahs.* New York: Henry Holt, 2007.

Napoli, Donna Jo. *The Wishing Club: A Story About Fractions.* New York: Henry Holt, 2007.

Pallotta, Jerry. *Apple Fractions.* New York: Scholastic, 2002.

On the Web

FactHound offers a safe, fun way to find Internet sites related to this book. All of the sites on FactHound have been researched by our staff.

Here's all you do:

Visit www.facthound.com

FactHound will fetch the best sites for you!

INDEX

adding
- big numbers, 22
- different number amounts, 20
- in order, 18
- left to right, 16
- small numbers, 22
- top to bottom, 16

addition, 10, 22, 27, 46, 54, 66, 71
difference, 36
digit, 100, 122
divided-by sign, 66
- definition, 72

dividend, 80, 81, 82
division bar, 84
division box, 84
division problem, 78, 80, 93
divisor, 80, 81, 82
even numbers
- adding, 104-105
- as cheers, 108-109
- as feet, 110-111
- as instruments, 106-107
- at a picnic, 112-113
- definition, 94

fractions
- comparing, 156-157
- definition, 138
- eighths, 150-151
- fourths, 148-149, 153
- halves, 146-147, 152
- sets, 154-155
- subdivided, 152-153
- thirds, 144-145, 154-155

groups, 124-125
minus sign
- definition, 28
- used as "take away," 34-35

multiplication, 54, 56, 66, 70, 71
multiplication table, 70, 71
number line, 49, 102-103, 126-127
odd numbers
- adding, 128-129
- as days of the week, 130-131
- as traffic signs, 132-133

definition, 116
plus sign
   definition, 6
   used as "and," 12
product, 56, 57, 58, 65, 71
quotient, 80, 81, 82, 90
remainder, 86, 87, 98, 120
sets
   arranged in different ways, 168-169
   as collections, 181
   combined, 174-175
   compared, 176-177
   definition, 160
   matching other objects, 166-167
story problems, 24, 40-41, 64, 90
subsets, 170-171, 172-173, 181
subtracting
   big numbers, 44-45
   left to right, 38

   small numbers, 42-43
   top to bottom, 38-39
subtraction, 34, 35, 44, 45, 78, 79, 89
sum, 14
symbol, 54, 76
tally, 124-125
times sign
   definition, 50
   used like "groups of," 62-63

TRISHA SPEED SHASKAN

When Trisha Speed Shaskan was a girl, she wanted to become a superhero. Her mother gave her a Wonder Woman costume. Her dad crafted her a tiara and bracelets out of metal to match. Trisha imagined she could fight evil, fly an invisible airplane, and get anyone to tell the truth. While she didn't grow up to be Wonder Woman, she still uses her imagination to write stories and to teach creative writing.

Trisha has taught creative writing to children and adults for 13 years. She has published 26 books for children, and more are forthcoming. She has an MFA in creative writing from Minnesota State University, Mankato.

Trisha currently lives in Minneapolis with her husband, Stephen, and their cat, Eartha, named after Eartha Kitt, famous for her role as Catwoman.

MARCIE ABOFF

Marcie Aboff is the author of picture books, early readers, chapter books, and magazine stories. She used to work as a feature writer for a daily newspaper in Escondido, California, and as a catalog copywriter for an advertising firm in New York City. She's written more than 20 titles for Picture Window Books.

Marcie loves visiting schools to talk to students about being an author, as well as helping them develop their own writing potential. When she's not writing or visiting schools, Marcie likes to play tennis, listen to music, see really good movies, travel, and eat as much chocolate as she can.

Marcie resides in Edison, New Jersey, with her two sons and one daughter. Their cat, Sneakers, also resides in the home, although he sometimes is under the impression he is the sole proprietor.

SARAH DILLARD

Sarah Dillard studied art at Wheaton College and illustration at Rhode Island School of Design. After living in Boston for several years, Sarah now lives on a mountain in Vermont with her husband and their dog. When she is not busy painting, she enjoys skiing and snowshoeing during the long winters, and hiking throughout the year.

FRANCESCA CARABELLI

Francesca Carabelli's mother says she was born with a pencil in her hand ... her left hand. She drew before she was able to speak. When she was 8 years old, they moved into a new home.

The hobby room in this new home was so white, that Francesca had an idea. She decided to make it more colorful by filling the walls with her characters. The ceiling was the only place that was not touched by her pencils. That was the beginning of her freelance career.

Since then, her funny characters have stayed with her everywhere, at grade school first, then at a University in Rome, where she studied Art and Literature. Now, the characters share the attic where she works, with her and her sweet black dog, Luna.